June 2013

John Hall, Author

Angela —
give this book to
whomever has been
diagnosed.
God Bless you
in your new works.

Beating Cancer Can Be Fun

Cancer Fighting Strategies for first time diagnosed cancer patients

John W. Hall

Author, Cancer survivor, M.F.T.,
Nutritional Consultant Intern(N.C.)

authorHOUSE®

AuthorHouse™
1663 Liberty Drive
Bloomington, IN 47403
www.authorhouse.com
Phone: 1-800-839-8640

First published by AuthorHouse 8/22/2011

ISBN: 978-1-4634-0363-8 (e)
ISBN: 978-1-4634-0362-1 (dj)
ISBN: 978-1-4634-0364-5 (sc)

Library of Congress Control Number: 2011908089

Printed in the United States of America

FOREWORD
BEATING CANCER CAN BE FUN (A HEAD START UPDATE FOR NEWLY DIAGNOSED CANCER PATIENTS)

This is the adventure story of a 72 year old man who tells of his Melanoma cancer recovery story in straight forward, candid, honest and sincere style with a purpose in mind for all people who become infected with any form of cancer. His is a message of hope for all those who have been told by their doctors that there is no hope for you to survive in your recovery process.

John Hall wants you to know there are two main types of medical practice, not one as The M.D.s and pharmaceutical companies would have you believe. Theirs is called the Allopathics. They do just three things: Cut out cancers, chemotherapy, and radiation. The chemicals they use are foreign to the human body and most often create violent and harmful reactions. The latter two kill healthy immune cells in the body at the same they try to isolate on cancerous cells.

The other medical practice is called Naturopathic. It includes natural plant foods, fruits, And waters. It feeds the human system with what is needed to heal. M.D.s as a group do Not talk about healing. It is almost a foreign language for them. As a result the F.D.A. will not allow any company to advertise products that CURE CANCER. Nor can doctors talk about a cure. Their hands are tied. They don't even have freedom of speech to talk about NUTRITION so that they can advise their patients before or after Surgery how to proceed to improve their internal health.

John's writings will inspire the reader to take charge of their cancer cells by conquering them with a nutritional schedule which will kill and eliminate them from the entire body. He learned his own recovery process from Dr. Ward Joiner, DC, HNT, my consultant who himself was told by his M.D. he would die of multiple causes and was a hopeless case. After John's removal of 10 metastasized Melanomas he sought out and found Joiner right in his hometown, Roseville, Ca. Joiner assisted John to analyze and interpret his blood results, with an eye to correcting and improving his number scores. The white blood cell count was very low, so was the NK(natural killer) cell count. John learned he was "almost dead." He decided to follow the exact recommendation of Dr.Joiner, taking supplements such as amino acid formulas, Thymus gland formulas, super enzyme supplements, alkalized waters to drink, green plant supplements, Wheat grass to crush and drink, and the myriad of alkalized foods. Since acids in the body build up on the blood vessels and cause toxic buildups, John learned of the importance of staying "alkalized." Why? Because cancers feed off of acids and cannot survive in an alkaline solution human body.

To this day John wonders why the Allopathics aren't teaching these important essentials to their patients? It is for this reason he strongly felt the need for this book. He wants others to know there are answers to the cancer quagmire. Though there are Still uncertainties as the cause of all cancers, there are remedies to prevent some Cancers and there are remedies to kill cancers in the bodies. "We don't have to wait for the medics to perform their long term studies to find a cure for cancer," says John. We who are survivors just need to share our stories of what we did to recover."

Moreover this book will talk of looking at our cancers and the "big picture". It will deemphasize the scare stories the medical profession insists we hang onto. It will provide specific recommendations as how to attack the cancers with proper herbs, greens, waters, supplements. The book is especially empowering to those who hang onto old myths about cancer and that it is a death sentence. It will put the patient in charge of when and how to use M.D.s and when to consult with a Naturopathic Doctor. Its main emphasis is on patients choice, not doctors preference. It main goal is to encourage ongoing education gained by the patient so that they will have hope and more energy in their bodies—to overcome the cancerous cells.

Lastly, there is a spiritual element to healing which John covers in the latter chapters. It includes his own philosophy on living and dying as well as the thoughts of others who deal with these issues on cancer in their professions or personal lives. The statistics on cancer indicate one of four Americans and their relatives will have to come to terms with proper nutrition, hearing the truths or falsehoods from advertisers, pharmaceutical companies, their doctors, and their ministers. Again, these are John's truths, his beliefs, and his sharings. Hopefully they will assist others to conquer their victim feelings, cure their bodies of cancers, and live an energy filled life until death do us part. Amen.

WRITTEN BY ALDEN OKIE, MASTER HERBALIST IN ROSEVILLE, CA.

(Note: I first met John just after his major surgery wherein the Oncologist removed ten metastasized cancers. John's immune system was in terrible condition and he needed a professional Holistic Nutritionist, so I referred him to Dr. Ward Joiner, DC, HNT)

ACKNOWLEDGEMENTS;

To my chosen one, Dr. Ward Joiner, DC, NC who provided me with hope, inspiration and the right nutritional protocol which saved my life.

To Alden Okie, sales consultant at Sunrise Natural Foods in Roseville, who guided my path just after the surgery, by referring me to Dr. Joiner.

To all the medical researchers and Nutritionists who provided excellent information and advice so that my body could heal from a deadly Melanoma. I salute you.

To my loving wife, Tina, who stayed faithfully by my side "for better or worse", especially when the light was dim and it looked like I would die soon.

To my son and daughters who gave a listening ear to me when I chose the Holistic path to recovery. Many thanks and much appreciated.

JOHN HALL, AUTHOR

TABLE OF CONTENTS

"It is true not everyone survives cancer with either the

Naturopathic or the Allopathic approach to healing. But taking

herbs, antioxidants, alkalized foods, green supplements have

zero side effects; as opposed to Chemotherapy or Radiation

which side effects are degrading, tortuous, painful and

potentially destructive."

- John Hall

CH 1
CANCER DOES NOT HAVE TO OWN YOU;
YOU CAN OWN IT

I am a recent cancer recovery patient, although I do not consider myself to be the traditional sort of "patient." You will notice there is a difference between how I felt when first notified I had melanocytes metastasized in my lymph system under my left arm,-- I felt like a victim, powerless. I was at the mercy of the scary cancer cells. And I was almost numb to the idea altogether. But I am happy to say I have entirely new outlook. No longer does cancer own me. I managed to survive and two years have passed since my surgery. Here is my story and the valuable lessons I learned during my recovery process.

Yes, first I allowed surgery to my underarm because that is how they have to detect melanoma cancer metatisizing. Doctors found 26 tumors and 10 metastsized. As I was waking up in the recovery room, the Oncologist basically notified my relatives that the odds for recovery were quite poor. When I awakened I could see the expression on their faces(he is a goner). A few days later my doctor explained there was nothing more he could do for me. There was no mention of alternative foods!

I began researching via the internet concerning Melanomas. I learned first: it's a fast moving cancer, meaning I could die quickly.

I had been feeling weak beginning 2008, now I felt even weaker, a victim at age 71. I thought maybe this was the price for getting old. Since I am a licensed Therapist, I had promised myself years ago never to allow myself to fall into a victim role.

Who was I to argue with the Doc at that time. He had the experience with others, and I sure did not want the cancer to spread. I signed a hold harmless paper for him, and he performed the surgery on 4-15-09. The prognosis for success with Melanomas in the lymph fluid is poor and I knew it. At that time I felt it was the right decision, however having done a lot of reading and research on the influence of surgeries on cancer cells, I have some doubts. Why? Some M.D.s have had experiences with patients wherein the surgeries caused the cancers to spread! This reminds me of a quote from Dr. Majid Ali, M.D. : "Cancer patients suffer twice: first with fear and suffering caused by their disease, and second from the ravages of a malignant system that forces not only toxic drugs of dubious value on frightened and gullible people(but also systems in surgery over which they have no control)."

My Oncologist advised me in his office a week later, in his non emotional manner: "melanoma could return to different parts of the body in 3 months, 6 months, or maybe never at all. We just don't know." Some people try Interferon treatments, he said, but in my case he did not feel it would benefit my condition: "the chemical has many negative, painful side effects over months, however with no ability to cure cancers." So effectively I was sent home to die. I did ask him that day whether he could refer me to a nutritionist, His response was"I just do cutting; we didn't study nutrition in medical school."

A few days later I remember from my past a 74 year old man in the early 90s told me his life was miraculously extended (since he took herbs for 5 years). His own Doctor was in disbelief and thought

5 years earlier he was presumed to die. He had a bad heart and lung condition, but an excellent Naturopath as his Holistic Doctor. My friend advised me to start taking herbs now—to maintain a healthy body. After that I did begin consuming herbs.

With that memory I visited the Sunrise Natural Foods store and sought out The best person to advise me what to eat for my Immune /System. The man I met there was Alden Okie, 15 years in the business of advising customers what to take for their particular malady. Alden shared his almost death experience of 6 years, when his doctor said he would die soon. Alden shared his path to recovery was all about learning what to eat and drink for his health. He recovered completely and works a full time schedule daily. He in turn referred me to Dr. Ward Joiner in Roseville, Ca., Holistic Nutritionist specializing in Immune Systems and Hormonal balancing.

Dr. Joiner shared his remarkable story of how he was sickly as a child and was disabled most years while growing up. Then later in mid life he became so sick his doc said he would die soon, to go home and get his things in order. As a practicing Chiropractor he was always focused on others health; now it was his turn. He said it was a blessing in the long run, as he began to improve his health via herbs, enzymes, amino acids, etc. I was quite inspired with his recovery because he had been diagnosed with lyme disease, chronic fatigue, lupus, and a very dysfunctional blood system. I hired him.

He immediately ordered a blood draw so he could interpret the functionality of my Immune system. It was low, low in many categories, hormonal balancing, NK cells and white blood cell count. Months later he told me: "you were almost dead when you first entered my office." He recommended a nutritional Protocol or plan to rebuild my Immune System so that my own body fluids and cells and organs could fight and kill the cancer cells. It was clear: The body can heal itself if it is fed correct foods, waters and supplement pills. Now for the first time I had hope. Now I felt I will be in charge of my life. I will have fun doing it because I am not a victim, not at the effect of the disease. And even if I die in the end, I will not suffer the devastating effects of the man made pharmaceuticals the medical profession is perpetrating on cancer patients: diarrhea, vomiting, pain and suffering over long term, oftentimes hospitalization; foreign chemicals inserted into their bodies which cause a significantly lowered immune system. In other words my chosen path to recovery would involve normal plants and foods and supplements-- with virtually no side effects. At last I felt no fear and I was confident I was on the right track.

I had studied and began taking herbs many years ago. I knew intuitively the body can cure itself from within, and not without with foreign chemicals. No matter how much pharmaceutical companies manufacture their drugs, it will never match the ability of our NK cells, T cells, IGF-1 proteins, our God given immune system. And now I was willing to accept full responsibility for causing my cancer, and to search out the remedies to cure the culprits and correct my diet in order to maintain optimum health.

What a great challenge! It is becoming more fun each day. It reminds me of playing sports as a youngster: you entered the game, after many practices, to Win the game; yet knowing you could lose.. Again, even if I were to lose, I learned there are no side effects to taking herbs, amino acids, enzymes, and alkalized waters: No pain but the potential for great gain. It can be a fun adventure. And as my Mom used to say: "Life is one great adventure."

The good news is: I was supposed to die 2 years ago, I did not. I am alive, thriving, carrying on with my life, sticking to my nutrition plan, and my blood tests are normal. I celebrate life daily, am grateful to all my friends and relatives who have been supportive for me, especially my wife, Tina.

To contact me, enter my web page via www.starlite@roguelink. us. You can also access my seminar presentation via Joiner's site: www.myholistichealth.net.

Before we move on to the next chapter I want to share the story of a woman, Maribel Lim (August 2003) about her pancreatic cancer amazing story. Its relevance to this whole Book will become obvious to the reader:

"Living with any serious disease can be difficult and challenging. I know how each one of you who has a serious ailment feels. I have also felt that way two years ago.

After reading my MRI results, my husband and I went from one doctor to another, to find out how to extend my life or cure the cancer. I cannot even count how many oncologists we went to. We were told this is an aggressive cancer with little chance of getting cured. Too little time is given for you to think, if you are still able to think straight given your condition.

'When I had the courage to ask the doctor how much longer I might live, he responded about six months. When I heard those words I felt the world standing still. Everything I heard the doctor saying was incomprehensible. I felt like a prisoner being handed a death sentence. I felt numb and the only thing I felt were tears running down my cheek.

The time the only thing I knew about was surgery. I braced myself for a 12-24 hour operation…my gall bladder removed…part of my liver, stomach and duodenum will be awaken away. If during the surgery they found a tumor which was too close to the pancreas, they will terminate the surgery and close the incision. Of course the other parts of my body will have been destroyed.

I consented to surgery even though it would be difficult. ...until the issue of blood came up. I am one of Jehova's Witnesses and as such we adhere to a Bible-based standard to avoid the use of blood, including a transfusion.

My doctors were indignant each time I refused a transfusion would be brought up...pressure, intimidation, threats. These, they resorted to, just so I would agree to a surgery using blood. Every consultation with doctors would result in depression, since it was not possible for me to change my mind. I would never compromise my position, even if it would mean losing my life.

This is the reason I was not operated on. Thanks to my belief in Jehovah's teaching. I would have been put throught Chemotherapy, the knife, or Cobalt and I would have been six feet under the ground.

So at that point my husband and I began researching. We read numerous books until we Finally came across "Alternative treatments." We learned that to treat a disease the whole body is involved. We learned that holistic treatment means to not just address the damaged part but the whole body. Each night we read a different book and we stayed up late just to learn about alternative treatments. Each research confirmed surgery was not the only solution, in fact, was not even required. I only had to change my lifestyle of eating foods. It was not easy and requires discipline.

In my case we started with 0 knowledge, with nothing. We were running up against time, very short, just six months. Every move must not be wasted. Every decision was Crucial and there was little room for error.

One very difficult aspect of having cancer is having "well meaning" people around you, who just want to be sympathetic and offer a little help or suggestion. Each one of them has an opinion to give, a little pressure here and there for you to try this or that, or just plain counsel on what to do. Of course I got confused, but I have learned not to be carried away by such pressures. The most important thing to remember when you get cancer is to stay focuses and not be swayed by mere talk.

I admit I lost confidence in what we were doing at times. Many times at nite the thought of not seeing the next day gripped me. When I had the painful attacks, I had the desire to undergo the surgery, but again, thinking about the blood issue, this firmed up my decision to go for the alternative treatment.

I took many food supplements; I would buy whatever they would give me. But I had a Stomach attack one evening and realized this wasn't enough. I felt a deep wound in my Stomach. I told my husband I thought my ending was close; then we prayed to Jehovah And hoped, perhaps, a herbal medicine might work.

The next day we met a Chinese sister in the faith. She introduced me to a cancer researcher who introduced Tian Xian, or China No11. We then met a Mr. Manuel Kiok who showed me the China I packet and told us about cancer treatment. For the first time I had hope my life might be extended. The first 6 weeks I also took China No6 and I recall black wastes coming out of my body. I knew then the medicine was being effective.

Six months passed and I am still alive. I was still weak and still uncertain. To measure the extent of my malignancy, I underwent an HCG test(Human Chorionic Gonadotropin) Test. The measurement was a number from the urine, to determine if my count was high or low. It was high in the beginning. I was overwhelmed and scared of that count. But over the next few months I ate better foods with my China I. I refrained from eating sugar, fat, salts, and oil, white flour, only organic foods. Each time I had an attack China I gave me relief.

I went off China I for awhile but then my scores went up again. So I went back on it, and finally my score fell to a safe range-no more cancer. After 2 years I am happy to inform everyone that my HCG count went from 80 to 51. I never thought I'd be alive after two years. Each time I go back to doctors now, they say I never had pancreatic cancer. I would explain to them I took herbal medicines that cured me. Invariably a shake of the head is the response I get. Here two years ago they told me I would die soon and they tried to rush me into surgery.

What, perhaps, could be the reason? Is it because I was weak then, and am strong now? Is it because they had no hand in my recovery? Or is it because they are reluctant to admit That alternative treatments work well? Of one thing I am sure, I'm glad I didn't follow their recommendations...

Lastly, to all cancer patients, I know, you too, can beat cancer. We can reverse cancer. Feed your body good nutrients through diet and supplements, thus providing your body the raw materials it needs to rebuild itself. Then feed your heart with the good feelings of Love, forgiveness, confidence in your abilities, a sense of purpose in your life, and a Trusting relationship with your Creator. I firmly believe we can beat cancer.

Thank you everyone!"

PERHAPS NOW YOU THE READER WILL EXPERIENCE THE DIRECTION THIS BOOK IS HEADED IN.

A sample of an herb in Costa Rica used as a medicinal plant. In Germany doctors prescribe herbs as part of their total treatment protocol.

John W. Hall

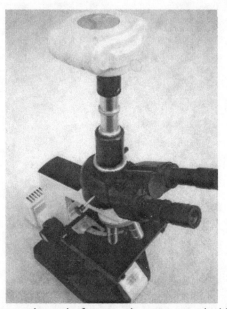

Darkfield Microscope shows dysfunctional nutrition in the blood, 12 months before it would show up under a regular microscope.

CH 2
THE BIG PICTURE:
WHAT WORKS TO HEAL THE BODY;
WHAT DOESN'T WORK

Albert Einstein posited "By academic Freedom I understand that the right to search for the truth and to publish and teach what one holds to be true implies also a duty: one must not conceal any part of what knows to be true"

I have evidence cancer can be cured, not just improved. My own case is one of recovering from Melanomas in the lymph system, which in most cases is suppose to kill a person within 6 months or less. But let us be clear on one thing. If we are talking about "miracle cures", that is not my understanding of what happens. Instead it is all about feeding the body with correct food and drinks so that the body's own natural defenses can kick in and do the work. It is the foods that build up the Immune System to make it stronger, not some injected chemical from the outside.

Yes, there are a few exceptions to using only natural ways: Chemotherapy for testicular cancer, for Juvenile Leukemia, and a few early detected Breast cancers. The bulk of cancer diagnosed patients, however, will die before the end of the 5th year.

My contention is: the new paradigm in America must be to get the cancer cells to stop dividing by repairing the NK cells, T cells, and IGF-1 proteins—all working in unison. Am I totally against conventional practices of the Allopathics? No but I believe patients must be able to choose between healing themselves from within, or outside in. And each person to be clear there are side effects from chemotherapy and radiation which can be tortuous—and there is no guarantee of recovery. And this freedom of choice implies each person diagnosed with cancer should be given complete information from both Naturopathic and Allopathic physicians prior to surgery.

It is also accurate that not everyone survives by using either approach. I want to emphasize that taking herbs, antioxidants fruits and veggies or supplements, there are virtually no side effects. Yet the side effects from chemo and radiation are degrading, tortuous, painful and potentially destructive, and they cause an attack on healthy immune cells, lowering a chance for a recovery. To support their position the Allopaths will cite testicular cancer cures and two others. Yet the bulk of their treatments clearly do not work at all.

A few more comments on the Freedom to chose: Freedom to choose is a great freedom. But if a patient has certain information withheld from him/her, is that really freedom ? Allopathic doctors in hospitals might say: Well the Naturopathics are quacks and they do no research to prove the efficacy of their plants and herbs. You know what I discovered in my research: The doctors and pharma companies don't want to know the answer. At the Federal level and through the larger non profit charities, they refuse to fund research programs to show the positive results of nutritional remedies. They intentionally block such a search!

From a spiritual or psychological perspective, if a paradigm in use, to wit, the Allopathic practices paralyzes us, freezes our brain to consider other alternative treatments, then the emotion of fear and hopelessness sets in, and we feel hopeless to seek out other ways to manage our cancer. If I were the Oncologist on any level III or IV cancer matter, I would feel quite frustrated, limited and disappointed to have to tell my patient: there is nothing more I can do for you. And if my system within which I operate prevents me from referring my patient for further evaluation and possible treatment, knowing my patient may die soon without another option, I would feel incompetent and irresponsible in this regard. Needed in these circumstances is some sort of Integrative Medicine and some intersharing between experienced Holistic Nutritionists and M.D.s

There is enough evidence around today that we know a good nutrition protocol can help people like myself to recover. As a Doctor then I would feel guilty not to help others heal in a paradigm wherein the medical profession might face these issues head on, instead of avoiding other alternate treatments. Of course a doctor can always say the person died from natural causes flowing from the cancer itself, or could blame someone else because the patient died sooner than later. Again, that is intentional avoidance, burying one's head in the sand. In my mind every physician in the U.S. needs to work on getting rid of whatever restrictive political legislation which disallows the use of nutritional protocols.They need to confront the FDC for their very narrow minded approach approach to treating cancers. The paradigm of using only chemo and radiation exclusively is almost over. The public is supportive of a change and so are many research minded physicians.

One thing for certain: there is no one chemical, food, or drink which will combat and cure cancers. Rather it is a combination of factors which will do the job within the body. We have natural made body cells which were created to stave off cancers: the T cells, NK cells, IGF-1 proteins, the hormone balancing in our system and all the organs—all functioning in an organized pattern of activity. My Holistic Doc, Dr. Ward Joiner, DC, HNT in Roseville, CA, assists cancer patients to look at their blood readings as a guide to improve the Immune System. He possesses the expertise, through his blood analyses, to recognize the deficiencies in the blood: white blood cells, testosterone amounts, NK cells, T cells, IGF-1, and best of all, to be able to recommend the finest quality supplements and foods. He also utilizes a live red blood cell analysis which can show a patient visually, under a super powered microscope, what a patient lacks in his body. This is analysis at a very deep cellular level. In my case it showed: i.e. not enough water, vitamin B deficient, hormone imbalance, any lack in digestion capability of an organ such as kidneys, liver function; and any lack of absorption within a particular body organ. From that information Joiner makes a recommendation of a natural food to correct the situation. This is a very scientific approach, not being used by many doctors in their practices. Remember doctors don't study nutrition in medical schools. Sadly this is further evidence of the lack of competence among M.D.s in their practice of medicine. What might benefit cancer patients, I suggest, is a coordination between M.D.s and the Holistic Nutritionists.

I have had several red blood cell analyses done because of my type of cancer which can spread so rapidly. They were a breath of fresh air for me. The reason: I can visually observe under microscope the expression of each cell to determine whether they are functioning well, or are dysfunctional and need some immediate remedy. Yes, I get to know if my Immune System is performing well or not. It eliminates a lot of fears and guess work for both the patient and the Holistic doctor. It goes way beyond the notion of hoping to get well. The red blood cell analysis is proof positive of what is occurring in your blood. This is truly Evidence-Based Nutrition at its best! Will it predict whether cancer will reoccur? No, but it will describe some precursors which might provoke further metastices.

In summary, like so many other recovering cancer people, I want some certitude we have a measurable chance for survival, with a minimum of suffering. The conventional approach to conquering cancer leaves one with too much guesswork, uncertainty, and suffering and pain are predictable. The Naturopathic way is much more predictable, not with absolute certainty, yet it lends hope to the cancer person that they can do something to move toward developing a healthy Immune system—and with no suffering. Then a realistic long term healing can occur, and the likelihood of cancer regression is not apt to occur. Why?

Simply a person has corrected his intake of foods and waters—which suppress or kill the DNA in cancer cells. Even if, in Breast Cancer as an example, a woman receives chemotherapy—and it returns later—if she hasn't been eating correctly, certainly healthy nutrition could be the remedy, isn't that right?

The photo of the Cat's Claw herbal medicine is a reminder of the chemicals within it which are anticancer, antioxidant. In the jungles of Brazil and Peru it is called the "Sacred herb of the Rain Forest." Current studies show it has positive results with boosting the immune system. Its active substances are alkaloids and tannins. It has also been used in the treatment of arthritis and rheumatism, diabetes and lupus.

CHEMO AND LANCE ARMSTRONG, RECORD BREAKING BIKE RACER

I would be remiss to not mention the story of Lance Armstrong's chemotherapy and subsequent suffering in order to overcome his

15

cancer. In his book "Its not about the Bike, My Journey Back to Life," Lance describes the truth about Chemotherapy better than anything I have read. These are summary quotes of his struggle to survive from testicular, lung and brain cancer:

"He(the doctor) would continue treating me with bleomycin, but his regime would be much more caustic than what Dr. Youman had :prescribed. He said: "you will crawl out of here...I am going to kill you. Evey day I am going to kill you and bring you back to life. We are going to hit you with chemo and then hit you again. You are not going to be able to walk. We are going to have to teach you how to walk after we are done."

Lance continued: "Because the treatments would leave me infertile, I would probablynever have kids. The Bleomycin would tear up my lungs and I would never be able to ride my bike again. The more he talked, the more I recoiled at my enfeeblement. I asked him why the treatment had to be so harsh. "You're worst case," he said. "But his is your only shot at this hospital." When Lance's male friend asked the doc about alternative treatments, he cut him off. The doc proceeded to tell him he would be better off at this Houston hospital than returning to Indianapolis to another recommended treatment.

Lance chose not to be treated at Houston, instead returned to Dr. Nichols who was the expert on testicular cancer in Indianapolis. He suggested a protocol of brain surgery first, then chemo. He said it was 50-50 but, if it all worked, Lance might be able to ride bike again. ..there was another protocol of platinum-based chemo, VIP(vinblastine, ifosamide, cisplatin), which was caustic in the short term, but which in the long term would not be so devastating to my lungs, as bleomycin. The doctor predicted he would more nausea and vomiting and short term discomfort. The doctor explained we don't want to do radiation to the brain because it has long term effects on the central nervous system; some patients end up with intellectual deterioration and cognitive and coordination problems. The doctor added there is a small risk of seizure after the brain surgery. Ultimately Lance chose the Indiana center for surgery and chemo.

In his Chapter 6 titled Chemo I want to share some details of what he had to undergo: "The question was, which would the chemo kill first: the cancer, or me? My life became one long IV drip, a sickening routine: If I wasn't in pain, I was vomiting, and if I wasn't vomiting I was thinking about what I had had, I was wondering when it would be over.

That's chemo for you. "…chemo was an endless series of specific horrors, until I began to think the cure was as bad as the disease itself. The effects of the treatment are loss of hair, a sickly pallor, and a wasting away…chemo was a burning in my veins, a matter of being slowly eaten away from the inside out by a destroying river of pollutants until I didn't have an eyelash left to bat. A continuous cough, hacking up black chunks of mysterious, tar-like matter from deep in my chest. Chemo was a constant, doubling over need to go to the bathroom.

To cope with it, I imagined I was coughing out the burned-up tumors. ..when I went to the bathroom I endured the acid sting in my groin by telling myself I was peeing out my dead cancer cells. I had no life other than chemo. I spent every winter and major holiday either on a Chemo cycle or recovering from one…I slept 10-12 hours a night, and when I was awake, I was in a funk that felt like a combination of jet lag and a hangover.

Chemo has a cumulative effect: I underwent four cycles in the space of three months, and toxins built up in my body with each phase…each time I would check into the medical center on a Monday and take 5 hours of chemo for 5 straight days. When I was not on chemo I was attached to a 24 hours IV drip of saline and a chemical to protect my immune system from the most toxic effects of the isosfamide, which is damaging to the kidneys and bone marrow. The final 2 phases cause me much pain and retching. The chemicals to treat the cancer came in plastic bags. The nurses had to wear special gloves to handle it, and on each one the label read "hazardous materials."

…Chemo doesn't just kill cancer cells—it kills healthy cells, too. It attacked my

Bone marrow, my muscle, my teeth and the linings of my throat and stomach, and left me open to all kinds of infection. My gums bled and I got sores in my mouth. And I lost my appetite, which was a potentially serious problem. Without enough protein, I wouldn't be able to rebuild tissue after chemo had eaten through my hair, skin and fingernails.

Chemo is a lonely process. My friend Och was there for me during the treatments. He knew has it brings ones spirits down; he had cared for his dad who died from cancer. He read the newspaper to me, and my mail and taught me how to play cards. He took me for walks around the hospital. Usually I would pass out as he entertained me. …I used to call my disease THE BASTARD…the tough part for me is I was losing my Independence and self-determination…oftentimes just looking at food made me vomit even harder. I was losing weight and then one day I noticed brown blotches on my skin.

Then I was sure my body was deteriorating … to add to my frustration, one of the big companies who sponsored me in bike racing threatened to cut out my contract. As I mentioned earlier I had no health insurance when first diagnosed, so now it had gone from bad to worse.

After several months of treatment Lance's blood readings began to improve. He finally went home to his mother's and tried biking. It was exhausting and several times he had to lie on the grass, thinking he might be dying. But he was determined to race professionally again. And he did. And won 7 Tour de Frances championships. As part of his story he did mention the importance of not drinking coffee, refrain from Red meats and cheeses and other things. Moreover, nutrition did play a part in his Recovery.

WHAT NEEDS EMPHASIS HERE IS LANCE MADE HIS CHOICE OF TREATMENT, AND MADE IT RESPONSIBLY. HE DID IT THE ALLOPATHIC WAY. NUTRITION-BASED COUNSEL WAS NOT PROVIDED HIM. THE QUESTION REMAINS: HOW WOULD A NUTRITION PROTOCOL, SIMILAR TO THE ONE I HAD, SERVED HIM? CERTAINLY IT WOULD HAVE INVOLVED MUCH LESS PAIN AND SUFFERING, AND WITH LESS RISK TO THE LONG TERM HEALING OF THE BODY.

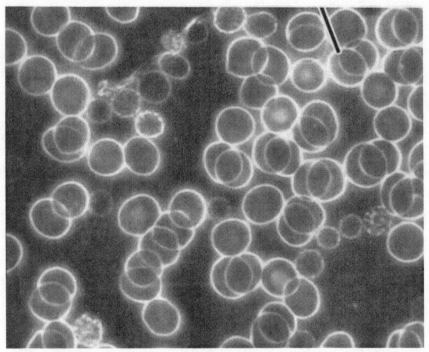

Live red blood cell analysis reveals a distortion in some red blood cells, which reflect a deficiency in vitamins, minerals, and water intake.

Cat's Claw Fights Cancer

"Immunity by definition (in medicine) is a condition in which

an organism can resist disease, or (in biology) a state of having

sufficient biological defenses to avoid infection, disease, or some

other biological invasion."

—*Wikipedia encyclopedia*

21

CH 3
GAIN KNOWLEDGE OF THE BODY
—ITS BLOOD—
AS NECESSARY STEPS TO HEALING

When I first met Dr. Joiner he repeated over and over again: "I don't treat cancer per: when we feed the cells, body organs, tissues and blood with correct foods and nutrition, the body can heal itself. It can utilize its own defense system against the free radicals (cancers) which are attacking it." It was just a few days after my surgery and I was in a very weakened state. He recommended a protocol of foods and supplements to enable my body to regain its strength. I could barely walk around the block back then. Today I can walk 3 miles and ride my bike for a half an hour.

The first month of ingesting the supplements I began researching the definition of the Immune System. I learned it is a bunch of structures and processes within an organism which protects against disease by identifying and killing pathogens and tumor cells. It protects against a wide variety of agents from viruses to parasitic worms. Detection is complicated as pathogens can evolve rapidly, producing adaptations that avoid the Immune system and allow the pathogens to invade the host.

The human system developed ways to neutralize pathogens. How? Simple bacterias possess enzyme systems that protect against viral systems. Some of these mechanisms include antimicrobial peptides called defensins or phagocytes. And the typical human system contains proteins, organs, tissues and cells that interact in a dynamic network. And over time this system learned to adapt to certain pathogens and learned an "acquired immunity."

Then what happens when disorders occur in the human system resulting in diseases such as auto immune diseases, inflammatory diseases and cancers? Here we are talking about a complex system so that diseases like AIDS, caused by tetrovirus HIV, can occur.

In contrast auto immune diseases—resulting from a hyperactive immune system—cause an attack by normal tissue as if they were foreign organisms. Some common diseases result, such as rheumatoid arthritis, diabetes mellitus type I, lupus, and thyroiditus.

Fortunately our blood samples tell the story of what is happening within our Immune System. It shows a weakened system if that is the case. Our NK (natural killer cells), T cells, and IGF-1 cells plus more will enable a Naturopath to read the results and suggest natural food remedies. In my case Dr. Joiner immediately recommended an amino acid complex for protein energy, an Adrenal formula, enzymes, antioxidant fruits and veggies.

As an experienced Holistic Practitioner, Dr. Joiner knew which natural food supplements my system was lacking. I was so blessed to have found him! Most folks with Melanomas IV are no longer on the Planet because the conventional Allopathic doctors are untrained or unskilled in dealing with most of the cancers.

After my fourth month of recovery I met Organic Jack of Newcastle, Ca.(in the hills just East of Sacramento), an Organic farmer of 30 years who explained to me and showed me the crops and how he protects them from disease without chemicals. He recommended I start on Wheatgrass and some organic vegetables. He suggested Wheatgrass has all the "live food" one needs to achieve good health. He attested many of his clients had recovered from cancer and other diseases and he showed me letters from them to prove it. I was skeptical at first and went home and did my research. A week later I called him and ordered some Wheatgrass. I still pick up my Wheatgrass each 2 weeks—as another means to assure good health. I attribute a good part of my recovery to organic vegetables, fruits and Wheatgrass.

I will mention here the photo on the front cover of this book: It shows a Neutrofil (yellow) engulfing an Anthrax bacteria (orange). This was taken with a Leo 1550 scanning electron Microscope, licensed under the Creative Commons Attribution (they do not endorse me, though I have clearance to use it in this book).It is the clearest evidence one can see on one segment of the Immune system at work! Yes, there is a real battle to be won: the strength of the Immune system defenses vs. the pathogen invaders! I, like many others, needed to visibly experience the reality of this dynamic to fully appreciate our own God given defense parts. If you recall, Anthrax is a deadly bacterial disease (Bacillus) which can be lethal if not counteracted, so the human body must be healthy enough to defeat the enemy. Eating Wheatgrass, antioxidants or organic fruits and vegetables is the way to assure the body is healthy. Jack lives it through farming without the use of pesticides or harmful additives. A la natural!

Another essential component to healing in my recovery is the importance of Exercise. I had no energy to exercise immediately after my surgery, then a few weeks later I began walking, slowly. Gradually my strength was returning. One of my coaches described how to breathe, so that by exhaling carbon dioxide, my body could release some of the toxins which had gathered in my organs and cells over the years. So now, 4 days a week, I regularly do weight lifting, a faster walking pace, running lightly on the machines, stretches and plenty of good alkalized water.

In the third month after the surgery, I was referred to a Detoxification specialist in Citrus Heights. He pumped 5 gallons of water through my digestive system(kidneys, liver, intestines)—to remove the heavy toxic buildup I had accumulated over the years. He emphasized it was essential , so the body could eliminate any blockages in those organs and cells. Otherwise my body would not be assimilate all the good foods I was eating.

All of the above mentioned components are essential to treat the Immune system. I am so pleased to have inherited the expertise of the workers in the health field who precede me. They indeed saved me from an earlier death and definitely from suffering from the cancer invasion.

"….dying patients should have the ultimate say as to what

treatments they should be able to take, and their choices should

include any experimental or unconventional treatments they

can get their hands on."

—*Richard Jaffe, Galileo's Lawyer*

CH 4
THE CHEMICAL ENVIRONMENT CANCERS LOVE TO LIVE IN

There has been much medical debate around which chemical environment does cancer live in-- to decide when and where to divide and multiply and become destroyers, We know now there are several sources which begin the process:

1. Poor diet is clearly one of them, In addition, Dr. Nicholas Gonzales feels we have a "cancer epidemic" in America from the depleted food value of crops, to processed foods marketed in our diets, from the many chemicals we breathe in, and we don't even know the full negative effects it is having on our bodies.

2. The increasing chemical contaminants in our air in the big cities

3. Radioactive Contamination:

    a. Also according to Dr. Nicholas Gonzalez, cell phones do cause contamination, especially involving brain cancer(p. 97 from the book "Knockout" by Suzanne Somers.

4. Electromagnetic Contamination:

    a. Comes from many sources: high wires, some factories,

x rays in hospitals, radio waves, microwaves, electrical

generators, induction machines and transformers.

The standards for which and how much of the impulses is healthy for a person is unclear. I have read many contradictory articles. Try reading the Governments safety guidelines; if you live close to an area of likely contamination then, via Internet, you will be able to compare their data with some of the "experts."

WHAT IS GOING ON INSIDE THE BODY FROM THE VARIOUS CONTAMINANTS?

It is now generally believed it is the toxic acidic buildup in our body organs, blood cells which triggers the cancers, resulting in metastises. Besides the standard blood lab analysis I have been having my Holistic Doctor perform the Live Red Blood Cell Analysis—to determine any dysfunctions at a deep cell level: liver digestion issues, lack of water in my body which can effect vitamin and mineral assimilation, a shortage of vitamins(I was recently low in vitamin B), insufficient protein intake. This test is done by taking a droplet of your blood, placing it under a super powered microscope and then showing a patient the photos of the different life forms on the computer screen. Most folks have never been analyzed at this deep cell level. I believe anyone who has cancer or early signs of cancer might consider this exam every few months, to determine whether body is functioning well or not. It is my belief this medical technology is highly underused to aid a patient in recovery.

Naturopaths believe strongly that a healthy diet is at the source of defeating cancers. Again my goal is to have a strong immune system as the only real solution to maintain good health and fending off cancer cells. As one example, Dr. Gonzalez cites the importance of accurate diagnosis: some cancer people need large doses of magnesium and potassium with large doses of calcium; whereas others need large amounts of calcium and don't do well on potassium/ In fact oftentimes they become depressed. The one thing he highlights is each form of cancer can be caused by our individual vulnerabilities to foods. He also cites there is also evidence building that having cell phones several inches from the brain for 8 hours daily is leading to a diagnosis of Glioblastoma, the worst form of brain cancer. Somehow the acidity in the body brings forth large tumors pressing on the brain. My sister in law recently had that diagnosis and passed away a few months after discovering it. Hers was an inoperable tumor.

To assure that a cancer diagnosed person strives to maintain a proper pH level appropriate to their body condition, I have listed a nutritional protocol in the next chapter which I use to bring my body into an alkalized state.

I monitor my acidity in the body with litmus paper 2-3 times per week. You can buy it in a natural food store and the outside of the bottle will show color codes(easy to read), to learn whether you are excessive acid or alkaline. The pH for Alakaline is a 7.0+ rating, for acid a 4.0-6.8 range. Consult with an experienced knowledgable Holistic Naturopath who can advise you as to a desirable acid-basic state for you.

It is also interesting to note that when cancer cells die they leave an acidic residue which is excessive. I take super enzyme supplements to wash them "down the sewer and out of the body." I have discovered cancers consider an alkalized state as their enemy because they love to live in an acidic environment. In my situation I am happy to share I feel better when I know my body is alkalized. I do worry when it becomes acidic, especially in a low pH range.

What have I learned about the importance of having an alkaline-acid balance in our system is: the kidneys and the lungs are the main filtering device to rid the body of excessive acids followed, to a lesser extent, by the skin. They are all engaged to facilitate preserving pH balance in the system. The kidneys help by eliminating solid acids, particularly sulfuric and uric into the urinary tract for elimination. However the kidney cannot effectively do this without bicarbonate and other alkalizing compounds, to neutralize the acid prior to its elimination. Chemically this is called buffering. Without this buffering with alkalized solutions, harsh concentrated acids with a pH of 4.5 or less would burn the delicate kidney tissues. If the kidney is flooded with too much acid and is in short supply of alkalization, this buildup of acid (we call acidosis) can set the stage for a wide variety of health issues, beginning with the disruption of proper cell function. Then, over time, numerous biochemical reactions become impaired and the stage is set for dysfunction and disease.

The lungs also work to keep the body's pH balanced by eliminating gas formed acids. Without describing the whole chemical process here, is that you can tell when the body is becoming overly acidic in an increased respiratory rate. The body is struggling to release excess carbonic acid from the system. If the body is over alkaline (alkalosis), the rate of breathing decreases. Again the body works hard to retain enough acids to maintain a proper acid balance.

The skin, through the sweat glands, is the final organ responsible for eliminating acids Through perspiration it flushes excess acids out of the system. It is noted that through sweating the body cannot eliminate acids as fast as through urinating. One way of determining excess acidity is the offensive odor which appears when we are overly acidic. In spite of this, continue exercising to eliminate toxic acids.

Let's now look at chronic low-grade acidosis—caused by a persistent build up of those acids in our system: looking at the dietary patterns of Americans today, there is an over consumption of acid-forming foods such as: proteins, grains, sugar, refined foods, coffee, and alcohol; and the under consumption of alkaline-forming fruits, vegetables, nuts, and seeds. This is what creates the heavy acid tilt in our systems today.

Next, how does this chronic acid condition harm the body? Mainly, the depletion of Alkaline reserves in the body puts a huge strain on the body, with negative side effects. The scientific studies I have read mention problems that will follow as the body fights for balance:

1. loss of calcium in the urine, the wearing of the bone, leading to osteoporosis
2. reduced bone formation
3. bones tend to become brittle and prone to fracture
4. the loss of potassium and magnesium stores in the body, with a tendency toward hypertension(high blood pressure) and inflammation
5. the breakdown of protein, results in muscle wasting
6. depressed protein metabolism, resulting in the inability to fully repair cells, tissues and organs
7. painful urination
8. damaged growth hormone
9. accelerated aging from accumulated acid products
10. increased production of free radicals—unstable molecules which cause cell damage—causing pain and inflammation—lowering immune capacity
11. greater oxidation of the free radicals and impaired activity of antioxidants substances that protect the body from free radical damage
12. tendency for connective tissue to weaken, causing inflammation and pain

13. decreases cellular ATP energy production—eventually leads to impaired organ function
14. encourages the spread of yeast and fungi, which thrive in an acid terrain
15. creates a more fertile breeding grounds for viruses, including HIV
16. reduces the size of the brain's energy reserve, causing weakened mental capacity
17. a decrease ability to perform exercise at a high level
18. due to increase in the mouth's acidity, leads to imbalanced bacteria effect, consequence of increased dental decay
19. a lessened liver detoxification, potentially causes a buildup of toxic residues in the body

This is a partial "laundry list" of the problems which can arise from faulty acid- Alkaline imbalance.

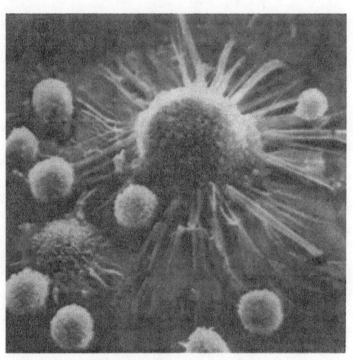

Cancer Cells being attacked by the immune system.

CH 5
THE pH FACTOR IS ESSENTIAL

It wasn't in my awareness about alkalized state vs acidic blood state when I was first being treated after my surgery. My blood tests were showing a reading of 4.5 pH extremely acid; so I immediately wanted to learn what to eat to improve it. I read everything in Internet, talked to Alden at the natural food store, learned about litmus paper (so that I could self test at home). I even tested the city water. It too was highly acidic. At the grocery store I discovered Fiji Water which tests 7.5 alkaline. Yes it comes from an island on Fiji. It is water that passes through volcanic ash aquifers to purify it. Later I learned about the Kangen machine which creates water at a 9.0 pH To this day I drink only alkalized water.

Below I prepared a list of foods to make one Alkaline and foods which tend to create acidity in the body vessels and organs. Remember cancers love to feed off of acid. Again to test oneself you can buy a litmus paper roll at your natural food store, take a small strip, placed under the tongue The color indicator will show it as yellow(acid) or green(alkaline). Optional: test the litmus with your urine.

THE FOLLOWING IS A PARTIAL LIST OF FOODS TO EAT AND FOODS TO AVOID—to maximize an appropriate acid-alkaline balance:

Alkaline producing foods   Acid producing foods

Watermelons red meats   beans fish

Lemons orange juice most grains   coffee   plums

Broccoli potatoes prunes   cranberries   eggs

Raspberries spinach gravy   distilled water
wine

Blueberries dandelion greens sour cream   f e r m e n t e d
food   and

Strawberries mineral supplements as digestive enzymes   a g e d
cheese

Pineapples K, Mg, Fe   (can   contribute   to   a   Mg
deficiency)

Bananas   alkaline water

Mineral water celery, artichokes, bok choy sausage, hot dogs, pastrami,
ham

lettuces, onions   yellow squash, zucchini,   peanuts,
walnuts, pecans

shitake mushrooms tomatoes, egg plant, organic fresh herbs

avacados, raw almonds, egg plant, sunflower seeds.

Organic Fish and Eggs   butter, milk, cream, ice
cream

organically farmed fish   sour   cream,
mayonnaise

white bass   super heated vegetable oils

pacific cod   barbeque sauces, mustards,

cold water prawns   caffeine, bakers yeast, brewers
yeast

tilapia, turbot   SWEETENRS:
                                          artificial   sweeteners,   barley

melt,

SWEETENERS brown sugar, corn syrup, glucose,
Stevia(not surgars)   fructose, sucrose, granulated
sugar
BEVERAGES
Fresh lemons in water
all Natural herbal teas

THE MAYO CLINICS DIET MANUAL ALSO INCLUDES A
NEUTRAL LIST OF FOODS:

Butter	corn
Margarine	tapioca
Sugar	coffee
Oils	tea
Syrup	
Honey	

It is noted this is only a partial list of foods (via internet see the
Price Pottinger Foundation ideal acid-alkaline balance of foods).

In conclusion, if a food tends to increase the acidity of the urine
after its ingested, it is classed as an acid producing food. If a food
increases the alkalinity (pH 7+) it is an alkaline forming food. The
effect foods have on urine pH may be quite different than the pH
of food itself, for example, lemon is highly acidic itself, but after
eaten it becomes alkaline (as the body metabolizes it to cause the
blood to become alkalized). If one has cancer it is important to learn
these distinctions, so you can know you are in control of the pH
factors. This is also true for other illnesses-- so consult your Holistic
Nutritionists who are knowledgeable in this area. In general the
medical profession won't tell you about the importance of staying
away from high acid producing foods. They prefer the public goes

on thinking their Allopathic way will cure or delay cancers rampage. Not so. This is why I personally don't trust the medical people as a group. They want to remain ignorant. They don't want to think outside the box of their own thinking. When they wake up it is usually when one of them is diagnosed with cancer. Then they will begin to seek the truth, as I did and am doing.

For more insight and understanding I suggest you read: THE ACID-ALKALINE FOOD GUIDE, a quick reference to foods and their effects on pH levels, by Dr. Susan E. Brown, PhD, CCN, a New York State Certified Nutritionist and researcher. Her coauthor is Larry Trivieri, Jr, a writer who coauthored a book: Alternative Medicine: The Definitive Guide. This book is the most complete guide to learn which foods will make your system acid producing, and which will help you maintain an alkaline state. The essence of this book is aptly stated on p.3, which I deeply adhere to :"Both health and disease begin in the cells, for it is at the cellular level that the vast majority of the body's multitude of interactions occur. For example, in order for the body's cells to function properly they need to receive life-giving nutrients and oxygen from the bloodstream and, at the same time, they need to release cellular wastes…only when the body is in an alkalized state can the body assimilate good nutrients and expel wastes. When the body becomes chronically acidic, however, these and many other cellular processes start to become impaired. Eventually, if acidity continues unchecked, the combination of a diminished oxygen supply to the cells and the buildup of wastes inside the cells sets into motion both fatigue and disease."

In summary, this was, perhaps, the most significant discovery since I got cancer: that CONTROLLING THE ALKALINE STATE IN OUR BODIES IS THE MAIN KEY TOWARD ENABLING THE HEALTHY CELLS TO KILL THE LITTLE ENEMIES. As my Holistic Dr. Joiner told me when I first met him: "I am not in the business of curing cancers. What seems to be the case is that a Healthy Immune system is like the old pac man game, just knocking cancers out one at a time." Of course he always added, in the final analysis it is the body which heals itself, no singular situation, chemical, or circumstance in itself. As he repeated: "We don't cure cancers, we help folks rebuild their healthy cells, strengthen them, maintain them—so that the body can heal itself."

Scientists have found eating broccoli with mustard can aid the vegetable's cancer-fighting ability.

Milkweed – A cancer fighting herb

"From a spiritual or psychological perspective , if a paradigm is in use(the allopathic approach) paralyzes us, freezes our brain to consider other alternative treatment, then the emotions of fear and hopelessness sets in—and we feel hopeless to seek out other ways to manage our cancer."

- John Hall

CH 6
HOW CANCER RESEARCH HAS FAILED

My relationship with M.D.s and the pharmaceutical industry is definitely ambivalent. I love them when they can help me; I hate them when I see how they do damage to peoples' Immune system, body cells, organs. During my first four months of recovery after the surgery, I attended a cancer seminar, mostly for women who had cancer surgery, chemotherapy and radiation. There were 43 women and 3 men in attendance at a major Sacramento hotel. The main speaker was a female Oncologist from Santa Barbara. After her talk she fielded questions from the audience. She spoke about the hopes that researchers might someday find the cure for cancer. She spoke about the recent "man made" drugs women were being injected with. Practically every question from the women was concerning: "How to I overcome the pain or discomfort from the drugs I am receiving?" Never once was there a mention of cure. It was all about how to gain relief from the burning, the chemical poisoning and the weakness they felt overall.

Worst of all there was no mention of nutrition in order to rebuild the immune system. So towards the end of the meeting I asked: "Dr., I notice there is no discussion of Nutrition to defeat cancer?" Abruptly, instead of sharing what she knew about some results of nutrition studies, she abruptly brushed the subject aside by saying: "There is no research monies provided to fund NUTRITIONAL research on cancers." I replied: "And why not?" She changed to another topic?

In crude lay jargon I felt the answer and avoidance on her part was a "cop out". It was deceptive, dishonest, and manipulative. It was designed to avoid any intelligent discussion of the plethora of research which has been done for 40-50 years on the good results of cancer defeating foods and antioxidants which help folks recover from their cancers worldwide.

Why do I react so strongly? Because there was no meaningful dialogue going on at the meeting that day. Like so many medical doctors she has narrowly defined an area of research. Like so many Allopathics she doesn't want to tell the folks the cold statistics provided by the National Cancer Institute: "Of all the patients who received chemotherapy or radiation, 95% will die by the end of the 5th year!" And it is well documented they will suffer pain and misery from the drugs during those years. Specifically, for women diagnosed with cancer and those treated with chemo and/or radiation, only 21% will recover after the 5th year. The other 79% will die before the end of the 5 th year.

To further reinforce my sentiments, three months after the surgery my Oncologist referred me to an Asian Oncologist Hematologist for consultation and possible referral to Univ of San Francisco Medical for their research program. Their program involved injections of chemical in an attempt to target and destroy cancer cells. I pointedly asked him: "are there any successes with Melanoma?" His reply: "Not yet." Are there negative side effects? "Yes", he replied. Needless to say I refused to accept any referral from him. He did admit, being from the Far East, that the present doctors in the U.S. don't allow any discussion of nutritional aspects of treating cancer—saying "our Union doesn't permit it." What he meant was their silent conspiracy will not allow it (since they have a lock on the monies received from cancer donations and from the huge profits M.D.s earn via their narrow definition of what might cure cancer, namely, chemo, radiation, or surgery).

Incidentally since then my Internet research reveals there are some M.D.s who have had the courage to step outside their "union", and they advise patients nutritionally for their cancer conditions and other auto immune diseases. Very few though. Yet still the bulk of research monies for cancers comes, not to fund alternative foods, but rather to continue down the same track which fails to truly help cancer patients defeat their disease.

Again, it is still incredible that after 50 years and billions of dollars contributed to the researchers in cancer that practically no progress has been made. We need some leadership and truth telling in Washington and amongst those who have worked so hard in the Naturopathic field. I would not be alive today if I had chosen the narrow and limited path of the Allopathic doctors and the companies who advise them. And it is also evident the medical graduate schools specializing in Oncology definitely need to be regulated by the government, since they refuse to self regulate their profession. They have walked free from having to use their Hippocratic Oath. They are not treating the whole man. They are not using an Integrative approach as a Model to represent their clients with the whole story on curing cancer. Yes, they studied a lot of chemistry in school, and then when they left they avoided the practical application of the subject when it comes to seeking remedies outside their narrow minded profession—as it applies to cancer research. It is the total Immune system which needs focus, not just firing bullets at isolated cancers. I am alive because my Holistic doctor and I are appropriately focused. I was abandoned by the Allopathics because they have no cure. They sell hope only, but it is veiled attempt to cloud the issue. We who have cancer deserve more. We need the best science can offer. Scientific investigation requires an open mind. And science today in the U.S. is intentionally avoiding the study of food nutrition, supplements, and waters with which to attack cancer cells.

DR. TULLIO SIMONCINI, ITALIAN ONCOLOGIST TELLS WHY HE BELIEVESALL CANCERS ARE FUNGI

It took a European Researcher and medical doctor to uncover the fact, according to him, the 100 year old hypothesis that has led science in circles is incorrect: scientists thought cancers were human cells multiplying without limits is an unproven theory that no one has proven yet. He contends there is no evidence at all for the genetic hypothesis and this gets proven out with the fact that orthodox cancer treatments do not work very well—especially when you look beyond the 5 year survival statistics. He says modern orthodox oncology is a failure and every doctor knows so in his heart and soul.

Simoncini: "My work is based on the conviction supported by many years of observations, comparisons and experiences, that the cause of the tumor is to be sought in the vast world of the fungi, the most adaptable, aggressive and evolved of micro-organisms known in nature." The aggression he refers to is so great as to allow it, with only a cellular ring made up of three units, to tighten in its grip, capture and kill its prey in a short time—in spite of a struggling prey. Fungus seems to be the logical candidate as a cause of neoplastic proliferation, he says. What describes yeast and fungi invasion perfectly is the metastatic cancer cells eat away through the protective barriers of an organ and overrun other tissues and organs. The reasoning here is fungi, during their life cycle, depend on other living beings, which must be exploited for their feedings.

The shape of the fungi is imposed in the environment in which the fungi develop. Fungi can modify their own metabolism in order to overcome the defense mechanisms of the host. Doctors know fungal infections are extremely contagious, and they go hand in hand with Leukemia 2 in children. If a child has had a bone marrow transplant and then gets infected with fungi, his chances for recovery are minimal. His chances for living with most antifungal agents are near 0. Of course we are talking about pharmaceutical agents, not the "miracle discoveries" of Dr. Simoncini with Sodium Bicarbonate as the remedy Fungi.

When Dr. A. Kaufman wrote his article on fungi and cancer a lady phoned into his Radio syndicated station talk show. Her three year old daughter had been diagnosed with Leukemia the year before, and the mother was convinced the antifungal drugs coupled with the natural immune system therapy, had saved her daughter's life. Another lady, having heard her story, had bone cancer and asked he doctor for a antifungal drug prescription. The results showed it killed the fungus and the cancer. She had told her doc the medication was for a yeast infection. However when she could no longer get the medicine, the cancer immediately grew back. Her doc thought that surely a few antifungal pills should have cured her yeast infection. Dr. Kaufman contended that the reason this medication worked was she did have a yeast infection, not a vaginal infection. He felt the fungal infection of the bone may have been mimicking bone cancer.

According to Dr. Kaufman, many cancer patients find the true fungal link to their Cancer, only to succumb to heart disease or immune deficiency caused by traditional Cancer treatment. "If this case were an isolated event, it might be referred to as 'coincidental.'

I have been able to plead with doctors of advanced cancer patients to at least try antifungal drugs for their cancer patients." Kaufman recommends a reading from the book: The Germ that Causes Cancer, to show that current cancer research is targeting the wrong cells. The whole point of his article is to show that cancer is a Fungus, and is not caused by a fungus.

Dr.Simoncini can be heard and seen demonstrating live fungal colonies and their destruction with sodium bicarbonate at http// video.google.it/videoplay. At the same site a breast cancer patient in Europe shares her success with bicarbonate. These videos reveal an astonishing truth about cancer and is safe successful treatment. Doctors and medical scientists have made the mistake of assuming that fungal conditions develop after cancer treatments have begun. Researchers contend that cancer therapies, aimed at destroying cancer, also destroy the immune system of the patient, leaving them vulnerable to yeasts and fungi, which multiply out of control. They consider these invading colonies to be "secondary" to the actual cancer. Again, orthodox cancer treatments are not working out well.

The essence of Dr. Simoncini's message is that the only solution to cancer, if it is a Fungus, is sodium bicarbonate. And, according to him, doctors are not very good at diagnosing fungal infections because their medical school training is based heavily on the role of bacteria and viruses in the area of infectious diseases. Fungi have been forgotten forever since the advent of antibiotics, and this is perhaps one of the biggest mistakes of Allopathic medicine. The overuse of antibiotics can lead to deadly fungal: infections: current tests for detecting the presence of fungi are both terribly scant and sorely antiquated. This is a serious problem for fungi are late stage infections that are provoking a wide range of life threatening diseases.

He adds that the anti-fungins that are currently on the market, in fact, do not have the ability to penetrate the masses, since they are conceived to act only at a stratified level of epithelial type. In order to achieve the most detrimental effect on the cancer tumors, the sodium bicarbonate must be put in direct contact with the damaged tissue. It is also possible, with the use of catheters, to insert the solution into the arteries; further drip infusions, irrigations and infiltrations could be used in places where the tumor has grown.

In addition, according to Dr. Parhatsathid Napatalung, bicarbonate physiology is entirely ignored in diabetes as it is in medicine in general. Who would stop long enough to think deeply enough to make the connections between acid producing diets (junk foods) and destruction of the pancreas? To quote the doctor: "The pancreas is killed if the body is metabolically acid—as it tries to maintain bicarbonates. Without the bicarbonates Insulin becomes a problem and hence diabetes becomes an issue. What flows from there is a chain reaction of inflammatory reactions throughout the body. The reactions would be included in the brain as the acidic conditions begin to prevail.

There are many causes of diabetes and cancers. Almost as many as all the heavy metals, Toxic chemicals and radiation contamination that will affect, weaken, and destroy pancreatic tissues. Bicarbonate physiology is foundational—meaning it forms the carpet that these poisons walk on. When the body is bicarbonate sufficient it is more capable of resisting the toxicity of chemical insults. That is why the Army suggests its use to protect the kidneys from radiological contamination. Much the same applies to Magnesium levels and iodine. They all protect us from the assault of noxious everyday chemicals we are subjected to in the air, food and water.

In conclusion, the research doctors are reminding us the over-acidification of the body is the single underlying most fundamental cause of all disease and then it is easy to understand bicarbonate physiology. I have used baking soda at different times and follow my acid-alkaline system frequently. It is one of my markers to know my cancers won't return. Also, as the economic crises explode forward, family finances are forcing more and more people to skimp on medications and physician visits and preventive screening—in order to pay bills. Orthodox thinking says that the more effective meds should cost more, the lesser ones don't work well. In fact baking soda is cheap at $2. a pound. And the use of a safer and very inexpensive Natural product just makes more sense than to spend thousands on lesser effective meds. This author cares about treating the masses with what works at a reasonable cost. Traditional medicine is bent towards serving the affluent, not the masses, in their pursuit of good health for patients.

D. SHARK CARTILAGE AS A POSSIBLE CURE FOR CANCER

Sharks are an amazing animal and we humans have had a relationship with this guy and gal for 500,000 years. Actually the shark has been on the planet longer than us, for 40 million years. Dr. William Lane wrote a book, published in 1991, titled "Sharks Don't Get Cancer—How Shark Cartilage Could Save Your Life." He relates in great detail, the Internal structure of sharks, their chemical makeup, and where and how they have survived over the centuries. The part that intrigued me the most is when he began talking about the Immune system of sharks.

Sharks have a strong and effective immune system. A shark's wounds heal rapidly, and Sharks are largely free of infections.

Antibodies contained in their blood successfully combat bacterial and viral infections. In addition they protect a vast array of chemicals that easily kill most mammals. Michael Sigel, who was chairman emeritus of the Dept. of Microbiology and immunology at the Univ. of South Caroline Medical School, was a Pioneer in the research of the immune system of sharks. What he discovered is that sharks, unlike humans, have an unusual immunoglobulin, a large amount of which is always circulating and ready to attack.

The sharks immune system produces antibodies against more than just bacteria, viruses, and chemicals. It also seems to help protect against cancer, whether the shark is bred and maintained in clean, open waters, or in carcinogen-laden closed contained waters. This is what inspired Dr. Lane to do extended research on the shark. He wanted to know how any data gathered could be related to a prevention or therapy for humans.

Research revealed sharks took in a lot of heavy metals into their system which might cause the flesh to be toxic. Because sharks are scavengers and take in the heavy metals, Mercury, nickel, copper, toxic in large quantities, but in trace quantities are good for our Health. The good news is those metals don't exist in the cartilage of the sharks. The reason: sharks don't have blood vessels so there is no way the metals can get into the cartilage. This is one reason the people in China have never suffered any side effects from the shark-fin(cartilage) in soups they have eaten for generations and love to eat. It was also noted at the time the women in China had a lower rate of breast cancer than in the U.S. He felt it was because the Chinese were consuming so much of the protein like fibers of cartilage in the shark-fin soups—this an important factor. These same fibers, when processed as shark cartilage powders can render the health benefits mentioned in his book.

More research has shown shark liver oil produces a huge amount of Vitamin A and, more than that, sharks are a "floating nutritional factory" that produce many benefits. Some of those include the enhancement of damaged human tissues in burns; also can act to help build white blood cells(and recently has been shown it can be used to protect against radiation). Shark cartilage has also been used in artificial skin surgeries, and in the form of capsules, powder, creams and suppositories for inflammatory diseases. Dr.Lane is also convinced that since shark cartilage is oftentimes more successful than the modern medicines used in cancer(which cause devastating side effects, more destructive than the diseases themselves), the shark benefits are destined to play an increasingly more important role in modern life.

Scientists at the Massachusetts Institute of Technology have discovered a substance in shark cartilage that slows the growth of new blood vessels toward solid tumors, thereby cutting off tumor growth, according to Warren Leary, Associated Press Science Writer. Progress in this area has been evolutionary, not revolutionary. Earlier studies in the U.S. in the 30s led to the finding in the 70s of a tumor necrosis factor(TNF) and later shown to damage the blood vessels that nourish tumors.The damage to the blood vessels reduced the flow of the blood and oxygen to the tumor cells, which then starve and die. In the late 60s Dr. Judah Folkman, M.D. of the Harvard Medical School, had been working on a hypothesis about the nature of tumors. A tumor is new tissue made of cells that grow in an uncontrolled manner. In normal tissue the growth is limited; the rate of cell reproduction is equal to the rate of cell death. Tumors can be benign or malignant. The benign are rarely fatal; the malignant ones proliferate and invade surrounding normal tissues, eventually metastasizing(spreading to other places in the body) via the blood and lymph system. A growth tumor mass (which is what this author had) will eventually suffocate vital organs, starving them of nutrients. When the vital organs die, so does the host. Stopping cancer then involves stopping these runaway duplication of cells. Dr. Folkman did pioneer work, with mice, to show that the blood supply carrying cancers need to be in a liquid medium to successfully build up the tumor. It took 20 more years before science realized what he coined as antiangiogenesis, the blocking of new blood vessels forming. Then tumor growth could be blocked.

Several researchers in the 70s experimented with mice and chicks and learned that Cartilage stopped the formation of carcinogenic tumors. This led to research with sharks. Dr. Carl Luer, Biochemist at Mote Marine Labs in Sarasota, Florida, was already doing research with sharks—because he knew, with rate exceptions, sharks rarely get cancer—either naturally or when exposed to massive amounts of carcinogenic chemicals. It was also learned that the protein inhibitors in sharks are much more potent and effective in preventing vascularization than other animals. Much research has been done with calf cartilage.

It is noted that Dr. Prudden at MIT worked with bovine cartilage for 20 years. He Successfully used the cartilage as angiogenesis inhibitors. Then it was learned that Shark cartilage is 1,000 times more effective in treating cancer patients.The best news: there was no evidence of toxicity; no abnormalities in kidneys and liver function or blood values. The Doctor reported a regression of cancerous tumors without the debilitating effects of chemotherapy, radiation or surgery. He concluded, after the FDA did the same testing, the cartilage had a major inhibitory effect on a whole range of cancer. According to Dr. Lane, in his book "Sharks Don't Get Cancer," no serious followup occurred in establishing cartilage as an inhibitor of cancers!

To his credit Dr. Lane continued his pursuit for further evidence of shark cartilage as a preventor of cancer and an inhibitor, by speaking to other highly acclaimed researchers both in the U.S. and outside in other countries. He became quite frustrated by the F.D.A. and FDC because they refused to follow up on the research and encourage doctors to use shark cartilage. They said it is a food and a natural substance, and they were only interested in chemicals made in a lab before they would approve it.What else is new? To this day, even though many survivors of cancer like myself can testify to the natural food and supplements we used to remain cancer free, the Feds still remain stuck in their basic position—even though hundreds of thousands of people die of cancers each year??

Without belaboring the point, research has continued with shark cartilage and is assisting patients to reduce their cancerous tumors. However Dr.Lane, to protect himself from being arrested, has had to encourage patients and doctors to continue using conventional treatments along with any shark powders or liquids on the market. He mentions the practicing Oncologists and physicians won't discuss the cartilage since it is a natural substance and is outside the narrowly defined realms of the conventional treatments and researches being done. Allopathic medicine is still trying to target cancer cells, but in the process seem to be torturing the patients in the process—and, again, most of them die by end of year 5.

"The second biggest killer in America is medical ignorance, and

it is the number

one reason people die"

-- Bill Fallon, founder of Life Extension.com

CH 7
CHOOSING M.D.S OR HOLISTIC NUTRITIONISTS(NATUROPATHS)

What first came to my mind, in considering how pharmaceutical companies and doctor approach their research on the causes of cancer, is that they unnecessarily set limits on themselves. In a sense they have gotten themselves into an attack mode to try to destroy cancer, in their traditional Allopathic thinking of utilizing laboratory made chemicals for their "chemotherapy"(is it really therapy?)and radiation(burning). They have numbed their minds to, possibly, other better alternatives.

There exists the option to explore all the plants, herbs, enzymes, vegetables, vitamins, and green supplements, such as mitake mushrooms, and on and on goes the list. remember this is how I and many others saved our lives. We chose the alternatives doctors refuse to explore.

We know now cancers cannot live in an alkaline body fluid system—and antioxidants fruits and greens kill free radicals. These are already proven well documented facts. I might mention here that high stress and anger do create a highly acid environment. And acids create acid buildup in vessel walls, preventing proper assimilation of vitamins, minerals and other necessary foods.

Chemo chemicals are de facto foreign to the human body. The body does not recognize them as supportive, but rather destructive. They kill healthy cells along with the cancer cells they attempt to target. In doing so they lower the immune systems ability to ward off cancer or kill cancer cell growth. Yet still billions of dollars are spent on the continuation of medical manipulations which do not work?

Such a waste! If we poured some of those dollars into healthy food research, certainly the end result would be improved—and much suffering in patients could be eliminated. Coincidentally the word Allopathic comes from the Greek word, pathos, which literally means suffering! Are present scientific methods leading to more suffering, or less suffering in patients?

And now to ask: Why not try natures way to knock out cancers? I did: to cure my cancers, by taking greens, antioxidants and the recommendations of my Holistic Doc. I have had no pain, no negative side effects, no discomfort by consuming natures products for my body. My blood readings of my progress are my evidence. With my permission you can visit Dr. Joiner's office where my records are stored; and notice I went from near death state to one of restored health in 6-9 months. I have fought hard to gain a 7.0+ alkaline pH reading, changing my eating habits from high sugars, high salts, too many carbs, red meats-- to foods which the body can pleasantly digest and assimilate. I also reduced some stresses by not hanging around people who are naysayers, who lock themselves into the notion that only the medical profession can save them. I am not saying the Allopathics have no value. What I am saying is they are failing to utilize the scientific methods of exploring all the avenues, with an eye to killing cancerous cells in the body. What this ultimately means is that to achieve the best results available they must look at nutritional feeding of cells --to boost the Immune System. The body, in a state of optimum health can and does heal itself. This has been undisputable for years.

So polarized at the other end of the spectrum, Allopathics are bringing about the bodies rejection of the chemicals they induce— resulting in a loss of hair, vomiting, nausea, diarrhea, pain, and extreme weakness What they can't get is that chemicals don't heal, and only the body can heal itself.

It is interesting to note that doctors or Oncologists invented a phase I-IV system to describe a patients negative retrogression with their specific cancer. This talk about how far along the cells have metastasized, invaded the body, or the downhill slope of losing life's battle, is(in therapeutic terms) negative thinking. It does not solve the problem of beating cancer. It is a doctors way of telling relatives and the patient of how hopeless the condition is. Instead of making alternative food recommendations, they talk from hope, to little hope, to no hope in phase IV. In other words the M.D.s don't know how to cure cancer and the questions arises: Do they really want to assist the patient to overcome a potentially fatal disease? It appears, then, the doctors just wait for the worst to occur, acting helpless and hopeless for their patients. Doctors as a group have become experts at describing negative events. They remind one of a war general who senses his army is being beaten in phase 3 or 4, and instead of using every means to defeat the enemy(cancer cells), they quit.

My Conclusion: doctors are not experts at utilizing every possible means to help the patients recover. In effect, the docs are opting for death for their patients, rather than researching for results Holistically or at least making referrals to the best of Holistic treaters.

The main point is patients deserve to know all the choices available to them so that they can elect alternative ways to beat their cancer. Why do doctors and pharma co's tend to take this narrow approach? The only answer today is it must be about money, profits, greed—which has cultivated an attitude of not caring about their dying patients.

Indeed if the rules of the game were changed, a paradigm shift, so that alternative treatments were extant—then attach money for their efforts—they would jump for joy to get onto this money train. They also would gain much greater job satisfaction when they experience more patients recovering, living longer, without so much pain and suffering.

Try to remember, doctors are not gods. They certainly don't have all the answers toward conquering our cancers. They are limited human beings with limited information. Recall when they were in medical school they were brainwashed to be in support of

Allopathic chemicals and procedures(which rarely work effectively, if at all).

My personal Oncologist, when asked by myself; "Why haven't you studied herbs, plants, or alternatives?"arrogantly replied: "I haven't got time. I'm too busy cutting on folks." By the way he did not offer me any hope for my recovery, and has never called to track my progress, after a year and a half. I could be dead by now and he wouldn't know it.

Suggestion: For those of us with cancer or an autoimmune disorder is to seek the best M.D. in your area, the best Holistic doctor or consultant, do your own research on Internet, Magazine articles, or books on the subject. The cancer took time to multiply so you can take your time making any decisions to act. Listen for any friendly advice or shared experiences from friends, relatives, testimonials from those of us who are survivors. Give yourself time to make an informed decision. Panic of any sort will not serve you. It is your body to serve and needs healing in the best ways possible. Lastly, become willing to live or die with your decision. That is true freedom. And your truth will set you free.

When I chose to inquire about natural remedies and met with Dr. Joiner, he spent a lot of time sharing his dramatic story of his recovery from many maladies. When questioning him about his past, and what led him to becoming a modern day holistic health practitioner, he humbly shared his life-long battle with many illnesses and health challenges. He said he vaguely recalls his early childhood days, as he was struck with high fever, diagnosed with rheumatic fever, and placed in the hospital. His symptoms were rapid heartbeat, tachycardia, withdrawn and weak. He was always comparing himself with youngsters his own age, since he was sick a lot and they were not. He felt separated from others and easily slipped into the "poor me' state of thinking.

As he got older he suffered from blood disorders, oxygen deprivation, fainting spells, relapses. Because of his health he was sent to his grandparents home, since he could not have fun with friends at home. He also suffered from speech impediments since birth and learning disorders from his illnesses, which tended to make him shy and quiet. His illnesses kept him home a lot and he was eventually home-schooled. Reflections on those years past now and the pain involved, helps him to understand them. He personally identifies with his patient's description of how they feel and their physical descriptions of their pain.

These are some of the reasons I hired him. He went on: At ages 12-14 his Naturopathic physician detected a faulty liver, so he placed him on vitamins and liver capsules. At 14 he was on the wrestling team, but 2 months later he became sick again and had to quit. He learned he had mononucleosis symptoms, leaving him weak. He was then placed on Jack Lalane's health program with vitamins, liver capsules—and he began to improve.

That is when he was sent to his Grandma's house. He had also mentioned his Dad had adult friends who were by today's standards "Intuitive healers." They guided him through high level meditations, self realizations, and teachings. For awhile this led him to become stronger as a teen and so he had more confidence jumping into adult life.

In 1991 his doctor told him if he didn't sell his Chiropractic business, he would die soon. He had developed several brain tumors and had bleeding ulcers. He did so and began his journey to discover the causes and remedies for his illnesses (for a complete story, read Dr. Joiner's testimonial on line at www.myholistichealth.net)

In summary, I was inspired by his personal experience with natural foods and his determination to help others get well through nutrition. In light of his many adult illnesses, he was finally placed on SSI Disability for 6 years. His desire to get well, coupled with his willingness to study and research the causes and possible cures for his illnesses finally led him to where he now works six ten hour days, six days a week, without fatiguing. He guided me from a near dead blood state to having energy within one-two months. Is he a miracle man? No but he does know what to works for a person, once he has their blood readings in his hand. The conventional doctors who give up on their chronic patients do not have this ability. Why? Because they are locked into a system which simply makes false promises to people and who refuse to look at the plethora of information about plants, herbs, enzymes, Adrenal natural formulas, Thymic gland formulas, and amino acids. It is interesting to me some of the so-called Integrative Medical models in residential programs are beginning, more and more, to implement nutritional schedules for their chronically ill patients; yet the private practice physicians are still into denial and ignorance of alternative nutritional protocols?

"Americans need to look at the statistics on deaths from cancers, then realize the money that has been donated to the American Cancer Society is being wasted.

After 50 years of failed research, by avoiding to also spend the money on researching the efficacy of natural plants and herbs and supplements—in an attempt to cure cancer—has cost cancer victims their lives, over and over again."

—John Hall, Author

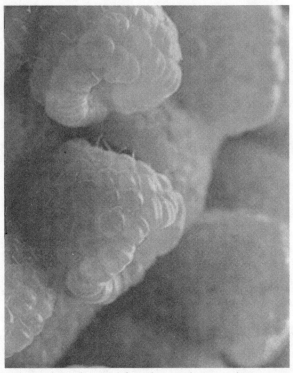

Ellagic acid is a phytochemical found is raspberries, also in strawberries, cranberries, pecans, walnuts – all have antioxidant properties, anti-inflammatory, and anti cancer.

CH 8
FUNDAMENTAL INFORMATION FOR THE CANCER PERSON

What was important to me when I was diagnosed with terminal Melanomas was where can I go to obtain sound advice and, mostly, what substances will work to prolong my life with a minimum amount of suffering? After I had located Dr. Ward Joiner, DC, HNT, and consulted with him, I had a burning desire to get the statistical facts on survival rates of Melanoma patients. Via the Internet I found the primary gatherers of statistical information nationwide is the NCI (National Cancer Institute) site. This shows how many people are diagnosed with cancer each year, the type of cancer, and how many died or survived. The statistic which caught my eye, which I will never forget, is: amongst those who had chemotherapy or radiation, 95% of them will have died by the end of the 5th year! And many will die before that. In other words, in reality, after 50 years of research and having spent billions of dollars researching remedies or cures for cancers, little progress has been made with achieving good results with patients. I am not saying to stop research, what I am emphasizing is that is biased, prejudicial work, and avoids spending a bulk of the money where there is success already in arresting or killing cancer cells. This is the main reason I chose the Naturopathic way—AND I SURVIVE AND AM HAPPY AND SUCCESSFUL AND ABLE TO TELL MY STORY.

One of the persons Suzanne Somers interviewed for her book "Knockout" was Bill Falloon, founder of Life Extension, Inc. when he said: "The second biggest killer in America is medical ignorance and it is the #1 reason people die." He enumerated, as an example, if you choose the mainstream or conventional treatment of cancer: cutting, chemotherapy, or radiation doctors are refusing to:

1. choose non toxic approaches to treating cancer
2. recognize that after surgery there is a higher risk of cancers spreading to other organs without a nutritional remedy
3. acknowledging surgery induces immune suppression
4. know it is important to boost as person's immune system before surgery— so that cancer cells that escape during surgery are killed by active immune cells.

The next question becomes if one is electing surgery, what is there to consume in order to inhibit cancer cells adhesion to normal cells? It is Bill Falloon's contention what works are two things: a supplement modified citrus pectin. In a study of Melanoma from a journal of the NCI, Melanomas were reduced by 90%! Secondly, another substance Cimetidine, suppressed cancer cell adhesion, according to the British Journal of Cancer 2002 edition. Again, he emphasized, better to use it before the surgery.

The mystery to me is why Oncologists are refusing to use these well researched advice. I learned from an M.D. recently that most medical schools don't teach Nutrition, or only take one class before they graduate.

Above all, SEEK DOCUMENTATION OF TESTIMONIALS FROM THOSE WHO OVERCAME CANCER after being diagnosed with Phase 2-4 chances for recovery— by having eaten a healthy change of foods, waters, antioxidants, enzymes and herbal supplements. Contrary to what many M.D. Oncologists might tell a patient, not only are Holistic Nutritionists not "quacks", but many patients have given positive, evidence proven information on how they recovered by maintaining their alkalinity, eating the right foods and avoiding acid producing foods. These patients are willing to tell their stories of recovery and, the best part of their stories, is that they did not suffer from the horrible side effects of chemotherapy and radiation for weeks and months.

In my own case my story is best summarized in a flyer I prepared for Organic Jack of Newcastle, Ca, the organic farmer with a heart and the knowledge to go with it: "I salute you, Jack! You have, along with Dr. Joiner, saved my life from a long bout of Melanoma in my lymph system. You convinced me to start juicing with Wheatgrass and I know this has been instrumental in my recovery from one of the most devastating cancers. According to my most recent P.E.T. scan(which can track malignancies), my body is now clear from all metastasized cancer cells. Allelulia!!"

Yes I do use other products, herbal foods, antioxidant fruits and veggie greens. I also take Joiners's recommended protocol of amino acids, vegetable enzymes, Adrenal supplements, Thymic gland formula, etc—to strengthen my Immune System. Once again, these natural foods don't damage healthy cells and don't cause suffering, as we know happens with chemo(therapy?) or radiation.

Lastly, Wheatgrass enabled my blood system to remain alkalized. Just recently I learned babies, born healthy, possess healthy Immune systems which is alkalized— and cancers cannot survive in alkalized state. They like an acidic state.

There are many testimonies available from those who recovered from their cancers via a nutritional plan: C.C., a middle aged woman recovered from a "terminal" Glioblastoma Multiforme (GBM IV brain tumor) diagnosis. Her PhD nutritionist placed her on an intensive protocol of diet, nutritional schedule, as follows: reduce sugar intake: Sugar suppresses the immune system and feeds cancer cells, along with omego-3 fats found in fish and flax, to slow tumor growth; Siberian ginseng, astralagus, cats claw, and mushroom extracts. Next she took 16 IP6 capsules along with genistein, bromelain, berberine, glutathione, quercetin and proanthocyanidins. (I have taken some of these supplements myself). This patient, C.C., also chose to refuse chemotherapy from her doctor, instead chose radiation. She followed the nutritional recommendations of her Nutritionist before and after the surgery, suffered no side effects as a result. Her Nutritionist explained:

"Even though many oncologists don't believe in taking antioxidants during surgery, 30 years of research reveals taking antioxidants during radiation and chemotherapy can be helpful.

An hour before her daily radiation treatments, patient took Vitamin C and E to protect healthy brain tissue and reduce swelling. Also included was shark liver oil, melatonin, St Johns Wort and whey protein. She summarized: Research suggests these supplements can maximize radiation's effect while protecting healthy tissue. As a result the patient had no complications or side effects from the radiation and the MRI done afterwards revealed the tumor had responded well.

One of the more inspiring testimonials comes from Dr. Lorraine Day, MD, who in 1992 was diagnosed with Stage IV Breast Cancer. She was chief of Orthopedic Surgery of San Francisco General Hospital, and for 15 years was on the faculty at U.C.S,F. School of

Medicine. As a medical doctor she chose not to have chemotherapy, instead choosing "Diet Therapy" as her Cancer Recovery Treatment program. She stopped eating animal products, instead eating only fruits, greens and vegetables. Dr. Day Completely Recovered from Cancer after her Diet Therapy program.

She had a lumpectomy of a small tumor, but the tumor soon recurred, became aggressive and grew rapidly. Yet Dr. Day rejected standard therapies because of their destructive effects and because these therapies often led to death. She chose instead to rebuild her immune system, using the natural, simple inexpensive therapies designed by God.—so that her body could heal itself. This too is my belief in my lifetime.

There are two videos of Dr. Day which I suggest you see On Line: "You Can't Improve on God" and "Cancer Doesn't Scare me Anymore." Her site: www. drday.com/tumor.htl

SOURCES OF BOOKS, MAGAZINES, JOURNALS TO FACILITATE YOUR CHOICE OF TREATMENT REMEDIES FOR HEALING YOUR IMMUNE SYSTEM

- Medical journals
- health magazines
- NCI data base
- Internet authors and their books on cancer recovery. See www.amazon.com/books
- books on the pH factor
- AMA(American Medical Association) results
- fightingcancerstrategies.com(an excellent resource)— possibly the best on the Internet
- My Personal Holistic Consultants:

Dr. Ward Joiner, DC, HNT Web site: www.myholistichealth.net

- Organic Jack—provides me with Wheatgrass on a regular basis

  Newcastle, Ca www.organicjack@sbcglobal.net

- Some of my favorite books include:

  ** "What are systemic enzymes and what do they do?" by Dr. Wm. Wong, PhD, ND, member World Sports Medicine Hall of Fame

  --he emphasizes the importance of supplementing our cancerous bodies with the following functions:

  -- a natural inflammatory

  -- anti fibrosis(scar tissue)

  -- blood cleansing, to break down dead material in the blood and reduce the thickness in the blood, and cleanses the FC receptors on the white blood cells—to help fight off infection

  ** "Knockout" by Suzanne Somers

  She interviews doctors who are successful in cancer recovery and Immune builders. These are some of the few medical doctors who have educated themselves to the Nutritional protocols and foods necessary for healing cancers. Excellent book.

  ** "Cancer Recovery Guide": 15 Alternatives and Complementary Strategies for Restoring Health" by Jonathon Chamberlain

  ** Outsmart Your Cancer" by Tanya Pierce

  An alternative non toxic approach to confronting cancer

  ** "Healing the Gerson Way": Defeating Cancer and other Chronic Diseases by Charlotte Gerson The focus is on nutrition for chronic degenerative diseases including cancer; a schedule for eating for a cancer patient; 60 complete recipes; 441 Pages(a large read), but worth reading as a reference book, describes coffee enemas as a way of detoxifying the liver and purging the body of toxins. This is a wonderful book for those who believe as I do: We as individuals and as a society have created our health condition; and we need to accept the challenge to correct it with good nutrients. Gerson claims her system cures cancers. I disagree to the extent that it is the rebuilding of the Immune System which enables the body to repair and heal itself. Of one thing she is certain: the medicalprofession treats the symptoms of cancer and are fishing up the wrong stream to prepare the body for a cure. A very good read.

** "The Acid-Alkaline Food Guide" by Susan E Brown an easy-to-read guide to the common foods that influence the body's pH level. an excellent guide for the cancer patient in preparing and cooking meals, or for ordering food from a restaurant. A short and inexpensive book.

• Last of all, the FDA(Food and Drug Administration)—specializes in preparing products for rare diseases, such as vaccines, biologics, blood research, radiation emitting products, and tobacco chemistry. Highly paid scientists, Chemists, and researchers making recommendations to doctors and consumers about the benefits and/or side effects of drugs. Their input becomes highly political for corporations, the Congress, and for all citizens. It is my opinion they are not the experts on cancer. Yes, they do have a corner on the market, in terms of which products will be approved and sold. They will not admit they have no cure for cancer and, in many cases, endanger the health of cancer persons.with chemicals that actually kill healthy immune cells—in an attempt to kill cancerous cells. I am a cancer survivor so I distrust and criticize much of the work they do. If I did what they promote, I would be dead.

- How To Prevent and Treat Cancer with Natural Medicine by Michael Murray This is a book which is very lengthy but rich in alternative treatments. Besides Diet it mentions studies on acupuncture, hydrotherapy and massage. He believes in complementing conventional therapy with Naturopathy, i.e. natural medicines and foods to aid in surgeries, chemotherapy and radiation. Endorsed by the Cancer Treatment Center of America, it is authored by 4 Naturopathic physicians. It provides some recipes for low salt, low fat diets which are useful. Again, they are hesitant to come on to speak poorly of the destructive nature of chemotherapy and radiation, not to mention the terrible suffering patients go through. Still a very good read, as a reference manual.

- www.pawpawresearch.com by Dr/ Jerry McLaughlin— PawPaw is an herb from a tree which, when processed and picked correctly, may be the best antioxidant in existence. Natures Sunshine Products prepares the standardized version of pawpaw. Be careful to research this one because there are companies selling this who have not prepared it correctly, and it might lose its effectiveness. I just began taking it, so if you know anyone with cancer, direct them to the research. It is noted that Dr. McLaughlin states Pawpaw actually can act synergistly with chemotherapy—making it more effective. He mentions though, with Juvenile Leukemia, chemotherapy does work and Pawpaw would not be needed in such cases.

I have many more suggestions to readers. Feel free to call me and I will share with you some of the quality products I have taken to overcome Melanoma. THESE ANTIOXIDANTS ARE FOR ALL TYPES OF CANCER, not just Melanomas.

John W.Hall, Author

530-355-8430

Best time to call is 7-9 am Pacific time, or evenings Pacific time

St. John's Wort – a more effective antidepressant than standard chemical
antidepressants

CH 9
WHEATGRASS AS A POWERFUL HEALER

My friend Organic Jack is an organic farmer who resides on his 5 acre farm in the Newcastle area of the Auburn, Ca, foothills. His labor of love is to continue growing Wheatgrass for folks—and he has evidence of many who have consumed his organic product and who have recovered from their chronic illnesses. First I will share his definition and description of this wonderful grass or juice; then two remarkable testimonials from his customers who were almost dead and yet who survived due to the many benefits of this organic 'miracle'.

Jack speaks: "Wheatgrass juice is literally condensed sunlight energy. It is one of the most potent healing agents on the planet. It is raw, living food full of rich, green chlorophyll."

Grasses are the foundational food for most plant based life. Wheatgrass was identified as the finest grass food of all after a series of intensive agriculture research studies headed by Dr. Charles Schnabel in the 1930s, 40s, and 50s. The study performed direct comparisons of wheatgrass against other well regarded vegetables, including broccoli, spinach and alfalfa. More research showed wheatgrass contains a broad spectrum of vitamins, minerals, antioxidants, amino acids, essential fatty acids and enzymes.

Juicing wheatgrass was started in the 1950s by Ann Wigmore independent of the above mentioned research. She was a Boston resident who suffered from a variety of ailments in her early life. She healed herself with wild weeds, herbs and greens. Her observations led her to conclude that wheatgrass was the best source of greens.

When you drink high quality wheatgrass juice, your body may produce spectacular results such as:

> **a physical and mental sense of well being
> **More energy and better sleep
> **Stronger immune system
> **Detoxification at a cellular level
> **lessened appetite cravings
> **Increased mental clarity
> **Steadier nerves
> **Improved eyesight and night vision
> **You suddenly start accomplishing more every day
> **visually seeing results via live red blood cell analysis, such as the unclumping of oxygen-carrying red blood cells

Recent research results show one fluid ounce of wheatgrass juice is equivalent to 2 and a ½ pounds of the choicest vegetables; is mineral rich containing 92 minerals needed by the body; is a complete protein containing 20+ amino acids; more than 30 enzymes found in the juice; it is 70% chlorophyll to oxygenate your body!

There has been much discussion about good health and body chemistry. And much talk about balancing body pH. For most, this means consuming more foods that produce alkalizing effects and less foods that create acid effects. Chemist have found this close relationship between toxicity (fermenting sugars), low oxygen states and corresponding sub-optimal performance of the body.

Since most of us develop an acidic chemistry over time (too much animal products), processed foods, we need alkalizing foods to restore pH balance. Wheatgrass juice is a powerful alkalizer It is also a preferred food because it seems to possess the ability to single-handedly change people's body chemistry when used in sufficient amounts. As a result, it is exactly the kind of food needed to jump start the processes of the body."

I have drunk wheatgrass for over 18 months, and it clearly works for me. I recommend anyone with cancer, or any other life threatening disease, use it regularly. You will notice a difference, believe me!

TESTIMONIALS FROM TWO OF JACK'S CLIENTS

1. Hi my name is Judy, I am 61 years old. In 2002 I was diagnosed with breast cancer. I had a lumpectomy, then had many months of chemo and radiation. My surgeon at UCSF said if I survived 5 years I would most likely remain cancer free. For the next 8 years I had regular checkups and mammograms. Then in Nov. 2008 I had an accident and went to a Chiropractor for the pain. Then I went for an MRI. To my horror, my breast cancer had returned and had metastasized to my bones, brain, lymph nodes and lungs. My body was a mess. How could this happen to me? I was not ready to leave my family. My Oncologist told me to get my affairs in order. She never even suggested Chemo so I knew I was in big trouble.

2. I have a cousin who had phase four breast cancer and she recovered, so I called her and asked her what to do. She had been free of cancer for 4 years. She referred me to the Bio- Medical Center in Tijuana, Mexico. When I went there they fed me with an herbal tonic and placed

me on all organic, raw foods diet. I decided to follow their instructions.

A few months later I was referred to Jack by my daughter. I told him my story. He insisted I start on Wheatgrass and organic foods. Within 3 months of following Jack plus the herbal tonic from Mexico, my cancer had improved and I felt better. I had CT scans a bone scans for a year and my lung cancer has now cleared up! Since improving, my Oncologist wants me to do more chemo. I said no. It took 21 months to improve my Immune System and I do not plan on destroying it. I feel great!

3. My name is Rick S.: "In 1995 I had a terrible cold sore. My doc said it was the worst infection he had ever treated. Then in 2006 I was diagnosed with Birkett's Lymphoma and given a 50-50 chance of survival if I followed the advice of my oncologist. These directions comprised exclusively a protocol of chemo in the hospital over 6 months. The last week, clinging to life in the ICU, I was released. I began a series of P.E.T. scans every 3 months. In December 2007 a scan came back positive with something deep in the chest. 3 months later that same area had grown and I was sent to a thoracic surgeon. The prognosis wasn't good. The lymphomas were between the pulmonary veins, heart and lungs and, since this was risky surgery, he postponed the surgeon to do more scans. I figured I was going to die. I felt sorry for my four year old daughter. I told a friend of mine and she referred me to Organic Jack. Jack told me if I would do 3 things, I might not die: 1. change to an all organic diet, second, increase my vitamins A,C, E with selenium, third, start drinking wheatgrass.

My choices were clear: give up and do nothing, or risk a small amount of money time and effort. I could die or risk being there for our daughter. I bought a juicer and several flats of wheatgrass from Jack, vitamin supplements, and I began juicing. On July 3 I called my doctor and he said: "I don't know what you did, you must have 14 lives because your cancer is clear." It's been 9 months now and no more cancer. I will close by saying the 10 years of cold sores I had have disappeared. Truly amazing! If you want to live longer, call up Jack and start juicing today. P.S. I truly believed if I had not met Jack a few months ago, my daughter, Paige, would not have a father today. Thanks Jack!

THESE TESTIMONIALS ARE EVIDENCE-BASED NUTRITIONAL STORIES. THEY ARE BASED ON WHAT WORKS, NOT ON FALSE PROMISES. THEY ARE PROOF ONE DOES NOT HAVE TO DESTROY HEALTHY IMMUNE CELLS TO BE CURED OF CANCER.

And notice the patients did not have to suffer.

"There are two primary choices in life; to accept conditions as

they exist, or accept the responsibility

for changing them."

—Dennis Waitley

CH 10
FEAR, DOUBT AND ANGER WILL
NOT SERVE A PATIENT WELL

YOUR EMOTIONAL RESPONSE TO YOUR CANCER COULD
PREVENT HEALING

It is true, if you don't quickly shift beyond your initial reaction of fear, doubt and anger concerning the fact you have cancer in your body, things could get worse. To not gain a proper perspective could bring about more pain, more conflicted relationships with your loved ones, and, possibly, hasten an even earlier death. I can, as a cancer survivor, relate to those feelings. When the Oncologist informed my relatives, after the surgery, of what he located and excised from under my arm (10 metastisized tumors) and then relayed the message to me. It was obvious he felt there was not much hope for my survival. I went immediately numb, like it wasn't real. Intellectually I heard the words, watched the cautious reactions of relatives as they spoke with me. But I was in denial it could be all that serious. After all I was healthy all my life and didn't think of myself as a sick guy.

A week later was a different emotional reality: I would die soon unless I took some action. At that moment all I wanted to do was find a Naturopath who knows about healing herbs. Fifteen years ago I had met a 74 year old man and he told me he had extended his life

6 years from a terrible heart condition, taking a protocol of vitamins and herbs. His doctor was dismayed. He said he made "a believer" out of his doctor, who previously did not trust nutritional approaches as workable. My friend convinced me to begin taking herbs for my health, and I did.

Moreover, armed with that past information and testimonial, I wanted to formulate a plan in an attempt to fight my Melanoma. I went to Sunrise Natural Food Store in Roseville, Ca. and spoke with Alden Okie, who had advised others about natural herbs and products for 15 years. He shared he had recovered from a chronic immune deficiency disease, after his doctor informed him he was a hopeless case and would die soon. Nine years later, with proper food and herb intake, Alden shared he is in perfect health, with normal blood readings, and is feeling great. Then he began giving me essential information on the immune system. At last I started feeling somewhat hopeful.

He then referred me Dr. Ward Joiner, DC, HNT, a Holistic consultant right here in Roseville. I gained an immediate appointment with Joiner: He shared his story with me so that I might be motivated to listen to his advice: He was sick and very weak as a child, became very sick again in his 30s with multiple diagnoses of several illnesses, followed by his doctor's prediction he would die soon. He chose to adopt a nutritional protocol which wiped out his infections, and today he is well and working 10 hours a day, helping others with life threatening problems. Inspired by the stories of these two men, I decided to hire Joiner to guide my path to recovery.

Dr. Joiner was very explicit with me: "I don't cure cancer, the body can do that provided it is fed correct foods." He recommended only natural foods and supplements, and no pharmaceutical manufactured products "because the body will not recognize them for its well being."

I provided Joiner with my recent blood analyses. He advised I was low-low in my testosterone, T cells, NK cells, and IGF-1. He told me later: "You were almost dead when you first walked into my office." Psychologically and emotionally, from the very first day I met him, I knew I would not have to suffer with my cancer, even though I might die. He felt by improving my Immune functions, my body might be able to fight off the cancer cells so I might recover. Over the months each time I visited his office and he showed me by blood test results, I had a feeling of optimism, not pessimism.

I could measure my forward progress by seeing my blood results and listening to his explanation of each blood indicator. Indeed I slept good that first couple of weeks.

MY OWN PSYCHOLOGY OF HEALING

As a trained Family Therapist in California; plus through my own personal life experiences; I had always wanted whole health for myself and my clients, not partial healing, but complete healing. Why? Because partial healing or no healing meant I would be in psychic pain— along with the physical pain. "No thanks", I said to myself (when cancer was detected).

I am not an advocate for suffering. I am anti-suffering for myself and my counseling clients. I said to myself: "Self, you could try the conventional methods of cancer treatment(chemo, radiation, interferon)—all of which induce a lowered immune system(our defense system); or the other choice, to elect a nutritional protocol or plan(which might work)—in order to build a stronger immune system. Dr. Joiner used these words as an analogy: "cancer is the enemy, remember the pac man game, we want to shoot them down one at a time."

Furthermore, my attitude before cancer was, and still is, if nutrition worked for others in their chronic illnesses, it might just work for me too. I knew the melanocytes would attack me and kill me over a few months. I therefore could not wait a year or more for the Allopathic long term longitudinal studies to originate. Through my own research I knew the narrow, tunnel vision approach of the Allopathics, oftentimes motivated by greed and power. I chose the naturopathic way. At least then I have a chance to live longer and or die with minimal suffering.

I rejected pharmaceutical pills, injections and began to avoid all food products which would destroy my immune system. Why? Because clearly the medical doctors admit they have no cure for cancer. And ultimately only the body's own chemistry can heal itself.

My emotional state today is free, peaceful and confident that I have made the decisions I am willing to live with. In that respect, I feel responsible for my outcomes. There will be no one to blame if my plan fails. True responsibility is a no blame state of being. It is clean and fresh of air. It is freedom at its best. The best word for my journey now is that IT IS FUN AND CHALLENGING, AND SO FAR I AM WINNING, going on two years since my surgery.

It is also fun since I get to share my victory with others who are having difficulty deciding their treatments of choice. If you were recently diagnosed with cancer, please just keep an open mind. Perhaps, too, a blending of conventional treatments combined with natures foods could become your choice. But for myself and with the knowledge I possess, the "quacks" are not the Holistic doctors, rather the M.D.s as a group who have their heads in the sand—when it comes to treating cancers. Until they are willing to adopt an Integrated Model of treatment, which includes educating themselves about nutritional remedies, they are not to be trusted.

They follow the mandates of the Federal government, who are biased in their approach to curing cancers. Little wonder that the death statistics for helping folks with cancer are so pathetic! Don't just believe me. Do your research and draw your own conclusions. Remember I recovered from the fatal disease in a phase IV Melanoma, and thank God I did not trust the establishment.

"The ultimate choice for a man, inasmuch as he is

driven to transcend himself, is to create or

destroy, to love or to hate."

—Eric Fromm(1900=1980), "The Sane Society"

CH 11
DEVELOPING A LISTENING TO HIGHER POWER, GOD, HOLY SPIRIT

HE CREATED US, HE CAN HEAL US

God gave us Life. We don't know how he did it. He provided us with all the water, trees, soils, plants and Sun so that we could be Healthy bodies on Planet Earth. He did not intend us to suffer. Just the opposite, to thrive and have vibrant energy. He also gave us the brain power to correct things as they go wrong. Since age 7, I have believed He is an all giving, all healing, all forgiving Spirit, and all Loving Creator. Many times in this life has he preserved my life: an auto accident at age 17 wherein the car I was sitting in hit a high wire pole and I was thrown through the windshield. Result was severe cuts and scarring and a giant concussion with headaches for 18 months.

Three years ago I was in an auto accident and, when they scanned me, an Abdominal Aortic Aneurism showed up. The surgeon saved me from bleeding to death by inserting a Stent in the large blood vessel, to straighten it out. This was truly a lease on life.

We all have stories on how we were protected from birth until the present by a power bigger than us. Why? I don/t know. What I do know is the Spirit-Creator is very generous with us, all giving. I believe He expects us to honor his creations, especially our own body.

If we are intentionally taking in poisons, breathing toxic substances, hanging around radioactive stuff, eating foods that are toxic and will cause cancer and other life threatening diseases, then it seems we are dishonoring His intentions for us.

My conclusions-- I eliminated my anger, doubts, and fears by doing the following:

1. Acknowledging the body was given all the necessary chemicals, hormones, blood cells, and organs to cure itself of cancer, if fed properly

2. Placing my trust in God, Creator of this body, and being appreciative of the modern diagnostics developed by our scientists, especially for the P.E.T. scan which enabled my Oncologist to specifically locate the malignant melanocytes.

3. I trusted the M.D. surgeon to help me make the choice to have surgery or not. I had 28 tumors, 10 metastisized, all removed in surgery

4. I trusted my Holistic Nutritionist, Dr. Ward Joiner, to become my coach. He recommended a protocol of supplements, foods, waters, greens, antioxidants, Thymic gland formulas, Adrenal formulas, and vegetable enzymes.

5. I took full responsibility, for better or worse, for my choice of the Allopathic way or the Naturopathic way

6. Having gratitude to my Creator for inspiring me to make good decisions
   One of my favorite quotes from the book "A Course in Miracles", says it the best: "... gratitude to God becomes the way to which He is remembered, for Love cannot be far behind a grateful heart and thankful mind." M55/58

7. Being thankful, to all of you who supported me during recovery.

8. Having gratitude, I reside in the U.S. where we have freedom of speech to elect a path to recovery

9. I chose some affirmations to overcome my fears and anxieties:

   "I am a creation of God and am worthy of good health"

   "I value life, with minimal or no pain or suffering, for myself and others"

   "I am open to Holy Spirit to provide the guidance I need to conquer my cancers"

10. I ask for enlightenment from on High that this disease, we call cancer, will be eradicated from the Planet soon

My feelings are strong and positive in spite of it all. I realize that "hope pills" the Allopathics tried to sell me, and the 'wishing you well" messages from friends are not enough to overcome or manage my cancer. My recovery requires some research on my part, concentration on sticking to a healthy diet, and my being open to any new, effective research results. It is my intention to remain totally responsible for whatever the outcome with these Melanomas. I have already extended my life years with the help of many people—and it is my hope whoever reads this book will achieve some of the positive results I am enjoying. And yes, as my Mom used to say: "LIFE IS FUN AND ONE BIG ADVENTURE!

John Hall, M.A., MFT, Author

CPSIA information can be obtained at www.ICGtesting.com
Printed in the USA
LVOW040116310712

292164LV00002B/25/P